A Story of Light

A Simple Exploration of the Creation and Dynamics of this Universe and Others

Pravir Malik, Ph.D.

ISBN: 978-0-9903574-2-1

To Sri Aurobindo

The journey to explore Light began in an unexpected place.

A Professor in South Africa, Dr. Dietmar Winzker, read a book of mine some years ago – Connecting Inner Power with Global Change: The Fractal Ladder. He wrote me a 16-page letter detailing his assessment of the book. To say the least I was thoroughly impressed that anyone would take the time to study the book in such detail, and then go on to very frankly lay out what they were in agreement with, and what not. Dr. Winzker's passion and commitment led to further dialog between us and in course of time he suggested I do a doctoral program from University of Pretoria in South Africa, where he had himself completed the first of his two doctoral degrees. He subsequently introduced me to Professor Dr. Leon Pretorius, his own Ph.D. supervisor, who was soon to also become my supervisor.

My research interest was in the area of Complex Adaptive Systems, and specifically in the mathematics of innovation of Complex Adaptive Systems. Simply saying, what captured my imagination was the question of whether a common mathematical framework, based on trends observed in time and subsequent insights into the structure of space, could explain the dynamics of innovation, regardless of the scale of the complex adaptive system, be it the size of a mere cell, or the entire universe.

Dr. Pretorius was an astute supervisor and led me through an essentially Socratic approach. He would ask question after question and let me discover my own answers. Since the field of Complex Adaptive Systems is essentially multi-disciplinary though, I found that I had to research a chain of adjacent areas in order to answer some of the questions posed. Pretty soon I had explored 25 areas ranging from Schrodinger's and Heisenberg's quantum equations, to the structure of the Periodic Table, to molecular plans in the human cell, to the stability of enduring civilizations, to the existence of dark matter and energy and their possible relation to the expansion and contraction of a universe.

In the course of this investigation the math of innovation in CAS had expanded to include over 225 equations, many of which were instances of application in the specific areas I had examined, and seeing that my enthusiasm to explore and connect seemingly distinct areas knew few bounds, Dr. Pretorius finally stopped me from exploring any more areas, especially since my thesis was fast becoming "daunting from a reviewer perspective" to paraphrase one of the potential examiners. The good news is that by the time he did this I had already happened on the mysteries of light.

Light, I found, was this extraordinary phenomenon, and in the mathematical framework that emerged held a foundational place such that everything in the universe could be thought to exist because of light. Light was the Holy Grail to arriving at a unified Theory of Everything, providing the clue to a Unified Field Equation as it were. Having glimpsed this I wanted to capture in as simple a language as possible the discoveries that continue to fill me with delight because of their simplicity in offering an interpretation for the creation and dynamics of a universe.

I was inspired by a non-technical book in plain English, written by Albert Einstein on his Theory of Special and General Relativity, and thought that if he could explain to a layman something as complex as Relativity, then I must be able to do the same on my interpretation of how everything emerges from Light. Hence I took it on my self to write this book that would not refer to a single equation, and would try and explain any complex phenomenon in simple terms. For those who know me, they will attest that this is a difficult task since it is easier for me to write in a more complex manner.

Necessarily this book is short with a key idea being more or less explained on a single page, and with the next key idea being dependent on the ideas already explored on previous pages. I have sought to cover the essential areas from my doctoral work, with the idea being that a reader with an excessively scientific or mathematical curiosity or mind could subsequently refer to the original dissertation if they so chose to. Details to access this appear later in the book.

Note that the nature of this work being more of that of the recognition and connection of pattern from area to area, there will undoubtedly be instances where the details I attribute to parts of such patterns may require further thought or correction. But I hope the reader may begin to glimpse something that I have glimpsed, and that these insights may be a constant source of joy and inspiration. If this is the case I further invite the reader to continue in creative-community with me and list some options for this at the end of the book.

In-Joy.

Pravir Malik,
San Francisco

What is Light? It is the great seed of the world and contains in itself all possibilities that are and will ever be. This first section sets the foundation for this claim, and looks at this in more detail.

- Light @c
- Light @Infinite
- The Big Bang
- What is Light?
- Time and Space structured by properties of Light

Everything is made from light.

But how can this be? How can the dense material objects that make up our world – the stones, cars, buildings – be made from light? How can our thoughts, our emotions, our cells and bodies be made from light? How can all the animals and plants and clouds be made from light? Here is a story that explores how.

To begin this story think for a minute as to how fast light is traveling. If we turn on a light switch in a room it immediately gets filled with light. If we turn on a light torch on a dark night it immediately lights up a path through the darkness. And yet the light from the sun takes about eight minutes to reach the earth. So while light is moving incredibly fast it is still not moving at the infinite speed we may think it is moving at. It is in fact, moving at a speed 'c': 186,000 miles per second in a vacuum.

Now, it is because the light is not traveling at an infinite speed that the world of matter can appear. All matter is made from atoms. So let us enter into the world of a hypothetical atom in the process of creation to explore how this might be. Imagine that some form of light originates from what will become the nucleus within an atom. That light would take some time, however small, to travel any minute distance within the forming atom. So for that light to express itself, to be noticed, it has to take some amount of a fraction of time. During that fraction of a time there is a build up of energy and it is this build up of energy that forms a packet or quanta, as it were, that allows matter to form. It is this build up of energy that variously expresses itself as quarks and leptons and bosons and subsequently atoms – the very building blocks of everything that is.

This build up of energy can only happen because light as it exists at speed c, can be thought of as translating or materializing some possibility or information that exists in light in its native state, that is, when light travels at an infinite speed. Hence the very notion of what is known as 'quanta' arises as a translation mechanism to further materialize information or possibility in light in its native state, in relation to other states such as when light is traveling at c.

While such quanta will not always result in the formation of matter, it always can be thought of as materializing some necessary information required to uphold the complex canvas of experienced reality. I mention this concept here, and leave it aside for now. Note though that I explore it in a lot more detail in subsequent books in the Cosmology of Light series.

But what else is implicit or made possible because of the finite speed of light? Well, imagine traveling on a ray of light from the sun to the earth. Imagine that you are in minute 4 of the approximately 8 minute journey. As you look back you will see that 4 minutes in the past you were at the sun. 4 minutes in the future you will be at the earth. And in the present moment you are somewhere between the sun and the earth. So this limited speed of light already creates the concept of time and specifically of the past, the present, and the future.

So, four incredibly fundamental things are created because of the finite speed of light: matter, the past, the present, and the future. It could be said, therefore, that this matter-based time-based universe is a result of the finite speed of light.

Light @Infinity

But what if light can be experienced at other speeds? What if light can travel at other finite or an infinite speed beyond what we experience it at?

Let us imagine for a minute that light can travel infinitely fast. Think about a big area or volume with a light source at the center. Now since the light travels infinitely fast it will fill up the entire volume instantly. This will be true no matter how large the volume is – it could be the entire solar system, an entire galaxy, or an entire universe. So that light is going to be instantaneously present everywhere – it is going to be omnipresent.

Now, since the light is already present everywhere in whatever volume, that is, since that light has already filled up the entire volume, there is nothing else that can arise there that is not of the nature of light. Even if something else were to arise, sooner or later it will be overpowered by light. So the light is all-powerful or omnipotent within that volume.

Further, since the light exists simultaneously everywhere in that space and has a complete knowledge of itself it therefore has a complete knowledge of that space or of anything that can arise in that space. So it is all-knowing or omniscient in that space.

Finally, it connects everything together instantaneously and holds these connections and the things connected, in its nature. So it is all-nurturing or omninurturing within that space.

11

So, as a result of the infinite speed of light, light appears to have properties of omnipresence, omnipotent, omniscience, and omninurturing.

The Big Bang

We hear about the Big Bang as the start of the universe. In this Big Bang matter is created. But we know now that the creation of matter is nothing other than the result of light traveling at the finite speed, and in this universe at 186,000 miles per second. So we can say that the Big Bang, the apparent start of the universe, is the result of a slowing down of light from an infinite speed to some finite speed.

When light slows down then energy accumulates in packets or quanta and this results in what was inexpressible being able to express itself as matter. This can be thought of in terms of an incredibly rapidly moving stream of water. If the water is traveling very fast then no boundary will be able to contain it and the energy will be continuous over the length of the stream. If it is traveling slower though, then the water will be able to be held by boundaries along the length of the stream. The energy in this case will be discontinuous and will appear in "packets".

We have arisen in the field of light that has a finite speed. It is difficult therefore to feel the reality of the omnipresent-omnipotent-omniscient-omninurturing light. We are attuned to perceiving in the material world that has resulted due to the finite speed of light. But if we could step back into the fullness of light in its pure state at infinity from there we would likely see that there can be different universes created as a result of light selectively slowing down to different speeds. The slow down to a different speed will create a particular kind of universe.

In this view the Big Bang is not an original creative event. It is likely a recurring creative event that allows something innate in light to express itself in time. But what is in light? What is in the omnipresent-omniscient-omnipotent-omninurturing nature of light? Everything we see around us, and an infinite more that we do not see, as we will discover.

What is Light?

When we think of light traveling at c, there are properties in the reality that emerge as a result of this. These properties as we have examined are of the

12

nature of matter, the past, the present, and the future. These properties are the result of light traveling at c and so can be thought of as emerging from light.

But what do these properties actually mean? And further, how do these properties relate to the apparently quite different properties of omnipresence, omnipotence, omniscience, omninurturing that can appear in a reality where the speed of light is infinite?

Let us consider first the properties of light that emerge when light is traveling at c: the past, the present, the future, and matter.

What is the past? It is the perceivable result of all the work and effort that has taken place so far. It is the foundation upon which the present and future will be built. It represents a status quo, stability, and even rigidity, and given that it is the result of the long play of time, it will not easily be persuaded to become another thing. It can be thought of as that which the eye can see when it looks around. There is "physicality" to what the eye can see and so the essence of the past is a kind of physical-ness. So, ingrained in light, is this ability to project or create physical-ness.

What is the present? It is the tremendous play of forces of all kinds vying to express themselves here and now. There is "vitality" that is present in this play and often it is the most energetic or forceful of the forces that will win out, as opposed to the most insightful or thoughtful. All the tremendous possibility of the future is seeking for expression now and so this essential vitality can also be thought of as a projection or possibility implicit in light.

What is the future? It is the inevitability of what will manifest. The great thoughts, the great ideas, the purpose, the possibilities will sooner or later express themselves in what we call the future. And the essence of this is thoughtfulness or a curiosity or a purpose that we can summarize as an essential "mentality". So embedded in light is this ability to project mentality.

And what is matter? It is the myriad crystallization into apparent diversity of the one essential reality of light, to allow for a play between these different sides or possibilities in an increasingly harmonious interaction. So its essence is "harmony", and this too can be thought of as an essential property projected from or implicit in light.

But what about when light is considered to move at an infinite speed? Then the properties that become apparent, as already explored, are those of omnipresence, omnipotence, omniscience, and omninurturing. But omnipresence has a physical-ness to it, omnipotence has a vitality to it, omniscience a mentality to it, and omninurturing a harmony to it. It may be said that physical-ness comes from omnipresence, vitality comes from omnipotence, mentality comes from omniscience, and matter comes from omninurturing.

So whether light is traveling at an infinite speed or the speed we know as c, there is something about the properties it projects, that in essence is the same. So let us refer to these essential properties made apparent through the worlds that are created, as Presence (from omnipresence), Power (from omnipotence), Knowledge (from omniscience), and Harmony (from omninurturing).

Time and Space Structured by Properties of Light

In considering the worlds that are created due to the way light moves or the play of light, we see properties that are projected because of it. In the finite world, the world that results from light traveling at the speed c, there is, relatively, smallness we can grasp and on which we can build. In the infinite world that results from light traveling at an infinite speed there is a vastness and fullness that is more difficult to grasp.

The notions of time and space are something entirely different in both worlds, and it could be said that 'space' allows the full play of everything meant by Power, Knowledge, Harmony, and Presence to be seeded in it, and that 'time' allows that seeding to flower into fuller forms with its passage.

In other words both space and time are not just abstract concepts but are essentially highly structured to allow physical-ness, vitality, mentality, and matter to become Presence, Power, Knowledge, and Harmony.

Space allows all the possibilities present in Presence, Power, Knowledge, and Harmony, and seeded in vast diversity, to evolve into more fullness through the time stages of physical-ness, vitality, mentality, and evolving matter.

Matter emerges and is made up of several layers that all are organized around the properties found in light. A way in which light has been studied is through the electromagnetic spectrum. So it makes sense that we should find the same properties of light – Presence, Power, Knowledge, and Harmony – present in the architecture of the electromagnetic spectrum.

But further, we will find that layers of matter – quantum particles, which include bosons, and atoms – are also structured or emerge along the same property-lines or property-families of light.

Electromagnetic Spectrum

The electromagnetic spectrum is a somewhat technical way to refer to light, essentially because amongst its range of properties it displays a tangible and simultaneous electric and magnetic or "electromagnetic" reality as well. So as light becomes more concrete to us one of the first forms it takes is as the electromagnetic spectrum. It must be the case that the previously surfaced properties of light – Presence, Power, Knowledge, and Harmony – exist in or even determine in some way what the electromagnetic spectrum is.

Let us look at this in more detail.

To begin with, we had already looked at how the speed of light sets up the nature of reality because of its speed. So for light traveling at c, past, present, future, and the notion of separation due to creation of islands of matter is the reality. It seems then that c architects the possibility of interaction in our system and can therefore be thought of a projection from the property of Harmony to create the basis for a matter-based harmony. So being, it can be suggested that the nature of the resultant interactions allows matter-based organizations regardless of scale, to come into their own, to grow into their boundaries, and to form bonds based on the sense of being separated from other perceived organizations. This notion of forming bonds seems also to be related to the "magnetic" in electromagnetic.

Further, the electromagnetic spectrum contains a range of waves - which enables many different applications we are familiar with, embedded in it. So for example there is a region in the electromagnetic spectrum that we call radio waves, and others that we know as microwave, infrared visible light, ultraviolet, x-rays and gamma rays that each make possible many technologies that we use every single day. These ranges essentially code a range of technological-possibility, and therefore the wave-range within the electromagnetic spectrum can be thought of as encoding or of expressing some kind of precipitation or projection or emergence of the property of Knowledge. Or put another way, the property of Knowledge that is found in light, emerges as the wave-range implicit in the electromagnetic spectrum.

The range of different wave-types or wavelengths implicit in the electromagnetic spectrum also moves with different frequencies. Since the speed of light is a constant, and it is known that speed is the product of frequency and wavelength, the greater the wavelength the lower will be the frequency of the wave-type. Conversely the less the wavelength the higher will be the frequency of the wave-type. And energy or power of a wave-type will depend on its frequency.

Hence, gamma rays that have a lower wavelength will have a higher frequency, and higher energy associated with it. Radio waves on the other hand that have a higher wavelength, will have a lower frequency and therefore lower energy associated with it. So implicit in the electromagnetic spectrum is a gradient of energy. But it is also known that the penetration power is dependent on frequency. Therefore, the higher the frequency or energy, the higher will be the power of the wave-type. Another way to say this is that the property of Power in Light emerges as the energy-Gradient in the electromagnetic spectrum. This Power aspect, by the way, seems to have been captured by the "electro" in electromagnetic.

If there is a large range of frequencies implicit in the electromagnetic spectrum then there is also the possibility of different types of masses implicit in the electromagnetic spectrum. Frequency determines energy, and mass and energy are related through Einstein's famous MC-squared equation. So pushing a little further it is not just that mass and energy are related, but a different kind of frequency or wave-type potentially perhaps allows a different type of mass to emerge. So the possibility of different types of mass seems to be related to the property of Presence in Light. In other words the property of Presence emerges as the possibility of different types of masses as suggested by the range of mass-possibilities that can emerge from the electromagnetic spectrum.

So we find that the four underlying properties of Light that we call Harmony, Knowledge, Power, and Presence are of the essence of the speed with which the electromagnetic spectrum moves, the wave-range within the electromagnetic spectrum, the energy-Gradient within the electromagnetic spectrum, and the mass-possibilities due to the electromagnetic spectrum, respectively. Further the description – electromagnetic – seems to have captured the Power-Harmony aspects implicit in light. In reality the electro-magnetic spectrum can likely be more completely described as wavearchetype-electro-magnetic-masspotential spectrum.

Quantum Particles

The electromagnetic spectrum or perhaps more accurately, the wavearchetype-electro-magnetic-masspotential spectrum, describes a first layer or translation of light into manifestation. This spectrum or field stands behind visible matter and perhaps it is fair to say that all matter emerges from such spectrums.

As already explored the finite speed of such spectrums allows the build of energies or quanta, and this in turn expresses itself as a series or as an array of quantum particles. The vast number of quantum particles that have so far been discovered has in turn yielded what is known as the Standard Model. This model is made up of what is called Quarks, Leptons, and Bosons, and a Higgs-Boson.

But if we look at these four fundamental quantum categories of particles from a property-based, or "functional" viewpoint a different chapter in the story of light emerges.

For instance, it can be seen first that the nucleus of an atom is made up of a combination of quarks. Specifically, a proton is composed of two "up" quarks and one "down" quark. Quarks have unusual names – up, down, charm, strange, top, bottom, with each subsequent pair belonging to a different "generation" of quarks. A neutron is composed of two "down" quarks and one "up" quark. Protons and neutrons together make up the nucleus of an atom. But we also know that the number of protons in the nucleus specifies the Atomic Number of an atom. Atomic number in turn uniquely identifies the element from the Periodic Table. Hence, an atomic number of 47, for example, specifies that the element is Silver. In other words it can be suggested that the unique properties of an element, the knowledge of what it is and how it will behave in the universe, is related to the quark. It may be suggested that quarks, therefore, are associated with the precipitation or emergence of Light's property of Knowledge in the quantum world.

When considering leptons it is useful to know that unlike quarks that only exist in composite arrangements with other quarks, leptons are solitary, point-like particles without internal structure. The best-known lepton is the electron. So the electron may be considered as a surrogate for the lepton class. The electron appears to be associated with the flow of energy and power. Further they appear to be the adventurers easily leaving the atom they are a part of. They also form locks or bonds with other atoms through the force of attraction and repulsion. In some sense they seem to be a representation or precipitation or emergence of Light's property of Power.

The bosons on the other hand are thought of as force-carriers. They are what allow all known matter particles to interact. The three fundamental bosons in this category are the photon, the W and Z bosons, and the gluon. The carrier particle of the electromagnetic force or spectrum is the photon. The carrier particle of the strong nuclear force that holds quarks together is the gluon. The carrier particle for the weak interactions, responsible for the decay of massive quarks and leptons into lighter quarks and leptons, are the W and Z bosons.

Bosons can be thought of as the precipitation of what creates relationship and harmony at the quantum level. Hence they can be thought of as the precipitation or emergence of Light's property of Harmony at the quantum level.

This leaves the other discovered fundamental particle the Higgs-Boson. In ordinary matter, most of the mass is contained in atoms, and the majority of the mass of an atom resides in the nucleus, made of protons and neutrons. Protons and neutrons are each made of three quarks. But it is the quarks that get their mass by interacting with what is known as the Higgs field. Hence the Higgs-Boson can be thought of as the mass-Giver. In other words it is what gives presence to the quarks and it can be thought of as the precipitation or emergence of Light's property of Presence at the quantum level. Just as there are multiple particles in each of the other 'families' it is likely that there will be multiple particles in the Higgs-Boson family. Recent research at CERN indicates that the Higgs-Boson may have a cousin.

So we see that again the underlying properties of light - Knowledge, Power, Harmony, and Presence - emerge as quarks, leptons, bosons, and Higgs-bosons respectively.

Bosons

The bosons as mentioned can be thought of as force-carriers and allow all known matter particles to interact.

But when we look at bosons in more detail there are three fundamental bosons – the photon, the W and Z bosons, the gluon - and one hypothetical boson – the graviton.

The photon is the carrier particle of the electromagnetic force. But in the scheme of things the electromagnetic force pervades everything and as explored in the section on the electromagnetic spectrum appears to be foundational to this reality we are in. So it could be said that it is related to Presence or is an emergence of Light's property of Presence at the level of quantum force-carriers.

The gluon is the carrier particle of what is known as the strong nuclear force and holds quarks together in their inherently composite arrangements. But we know that quarks are related to or are an emergence of Knowledge at the quantum-particle level. Hence it must be that the gluon is an emergence of Light's property of Knowledge at the level of quantum force-carriers.

The W and Z bosons are the carrier particle for the weak interactions, responsible for the decay of massive quarks and leptons into lighter quarks and leptons. But this usually is accompanied by the release of energy and power and so it must be that W and Z bosons are an emergence of Light's property of Power at the level of quantum force-carriers.

This leaves the hypothetical graviton that is thought to be the carrier particle for the force of gravity. But gravity is what holds astronomical objects together in relationship. Hence it must be that gravitons are an emergence of Light's property of Harmony at the level of quantum force-carriers.

So we see that the underlying properties of light - Presence, Knowledge, Power, and Harmony - emerge as photons, gluons, W and Z bosons, and the hypothetical graviton at the quantum force-carrier level.

Atoms

Quantum particles in turn create atoms, and all known atoms can be classified into one of four groups: The s-Group, p-Group, d-Group, and f-Group. But what are these groups and are they too emergences of Light as it continues its journey of crystalizing the possibility or potentiality within it?

s-Group

The s-Group consists primarily of what are known as alkali metals and alkali earth metals. These alkali metals are known to easily lose electrons and form what is known as positive ions. When they lose electrons energy is gained, but when the electrons are taken up by other atoms in proximity there is a lot of energy released. Some have referred to these groups as "violent worlds" and it has been pointed out that stars shine because they are transmuting vast amounts of hydrogen into helium, both of which are s-Group elements. So one gets the sense that the s-Group may be an emergence of Light's property of Power.

Stars and suns are also known to be the crucibles where all the different kinds of atoms are created. So these furnaces of power by virtue of their heat and high pressure are able to force electrons and protons and neutrons to come together to create all the different types of atoms known in the universe.

But philosophically what are the s-Group atoms or elements? These are atoms where there is an equal likelihood of an electron being anywhere in a symmetrical sphere around the nucleus. By the way, as will be seen, the other groups are all similarly defined by likelihoods of electrons being within a possible pattern around the nucleus. The patterns that distinguishes s-Group atoms is a sphere, and since all other patterns can be thought of as occurring within the sphere, in some sense this is like an imprint or precipitation or emergence that allows other kinds of emergences to surface within it. So the elements that are part of the s-Group may be thought of as the adventurers with courage who venture into a brave new world to create some foundation by which all other element-creations can follow.

The fact that hydrogen and helium are known to constitute 98% of the Universe relative to other elements therefore makes sense in this view, especially since hydrogen and helium provide the fuel with which the star-furnaces manufacture all other elements.

So s-Group elements seem to embody functions such as power, energy, adventure, courage, and can be thought of as an emergence of the property of Light to do with Power.

p-Group

Atoms or elements belonging to the p-Group are those with the likelihood of electrons occurring equally on either side of the nucleus, like a dumbbell of sorts.

There are some very significant elements in this group that are part of the metal, metalloid, non-metal, halogen, and noble gas sub-groupings. Carbon, Nitrogen, Oxygen, and Silicon are some of the sample elements. Looking at the types of elements present in this group it is as though all the element possibilities have been represented within it. It is perhaps that the possibility of ideas behind all elements has emerged in this group and one can hypothesize that this group may be a reflection of the property of Knowledge, forming archetypes from which all other elements are created.

Philosophically, the one spherical shell or probability cloud (s) becoming two shells like a dumbbell (p) signifies the creation of an essential polarity within a unit space. So in a sense the one becoming two is the first instance of variability in space. The two, spaced in 3-dimensions around the nucleus in a sense create six switches becoming an attractor or allowing a vaster number of different kinds of elements to surface. So there is a sense of the 'idea' of the element that comes into focus.

But further, the essential elements that allow both thinking and virtual thinking machines to come into being, are also contained within this group. Carbon is the basis of DNA and of all life. The fact that Silicon, directly below it in the Periodic Table and therefore sharing essential qualities, is considered the basis of all virtual thinking machines is therefore perhaps significant and may reinforce the notion that the p-Group is a precipitation of Light's property of Knowledge.

d-Group

The d-Group comprises the Transition Metals. These metals are generally hard and strong, exhibit corrosive resistance, and can be thought of as workhorse elements. Many industrial and well-known elements sit in this group: Titanium, Chromium, Manganese, Iron, Cobalt, Nickel, Copper, Zinc, Silver, Platinum, and Gold, amongst others.

The elements and atoms in the d-Group are equally likely to show up in four possible lobes or probability-spaces around the nucleus. Four lobes occurring in multiple possible planes around the nucleus will likely create a space of stability, since there is a possibility of four lobes creating the four vertices of a tetrahedron that has been positioned as one of the most stable shapes in the universe.

Much of the constructed world around us is created from these elements. Further, most of the series in the group easily lose one or more electrons thereby easily combining with other atoms to form a vast array of compounds. Also, looking more broadly at the function of these elements, it can be seen that these metals exist for service, to help bring about perfection in the constructed world, to help much of the machinery in which they are used, and to assist the processes dependent on them to be completed with diligence. Hence, these transition metals appear to be an emergence of Light's property of Presence.

f-Group

The f-Group comprises of the Lanthanides and Actinides. Philosophically, elements in the f-Group consist of 6 lobes around the nucleus within which an electron may be found. 6 lobes will exist in multiple planes around the nucleus and suggests the notion of extended relationship and collectivity: the attempt to build larger and larger bonds within a small space. Considering this it is likely that the f-Group is an emergence of Light's property of Harmony.

Thinking about Lanthanides, some interesting facts may reinforce this notion. First, the spin of electrons in a lanthanides' outer shell is aligned, creating a strong magnetic field. The notion of creating a strong magnetic field seems to be consistent with the notion of engendering a collectivity through the ordered attraction and repulsion of elements. Second, these elements curiously occur together in nature often in the same ores and are chemically interchangeable also suggesting the notion of forming a tight intra-group collectivity.

Actinides on the other hand are inherently radioactive. This implies that these elements have inherently crossed a threshold of stability and have the urge, over their own half-lives, to decompose into other elements. This natural urge may suggest some boundary conditions on the notion of collectivity and nurturing, giving additional insight into the nature of collectivity and nurturing. Further, the entire actinide group, as opposed to the lanthanide group that is inherently stable, is unstable. It is curious that both these should be part of the f-Group, and they must provide insight into boundary conditions into the notion of collectivity in elements.

LIFE: EXPRESSION OF LIGHT

While the notion of life is complex, things associated with life – such as living cells, love, sensations, urges, feelings, and thoughts, can all be seen as being related to Light's properties of knowledge, presence, power, and harmony.

Cells consist of molecular plans that are embodiments of the properties of light. The philosophical and even practical meaning of love becomes clearer when we consider the layers of light that have emerged to create matter, and living cells. Sensations, urges, feelings, thought that living beings are usually endowed with can be seen as emergences of Light's properties of Presence, Power, Harmony, and Knowledge respectively.

Cells

In its journey to create a universe, light has projected aspects of itself to structure the layers of the electromagnetic field, quantum particles, and atoms. In so doing it has allowed all of matter, and all arrangement of molecules and so even the basis of life to emerge.

But let us see how the properties of light structure the living cell.

As is known, every living thing on Earth uses a similar set of molecules to eat, to breathe, to move, and to reproduce. There are molecular machines that do the myriad things that distinguish living organisms that are identical in all living cells. This nanoscale machinery of cells uses four basic molecular plans with unique chemical personalities: nucleic acids, proteins, lipids, and polysaccharides.

Let us take a deeper look into the personality of each of these molecular plans.

Nucleic acids basically encode information. They store and transmit the genome, the hereditary information needed to keep the cell alive. They function as the cell's librarians and contain information on how to make proteins and when to make them. They are hence, the keepers of a cell's knowledge, its wisdom, its ability to make laws, the vehicle to spread knowledge within cells and to the next generation of cells. Being so, we can see that nucleic acids can be thought of as an emergence of Light's property of Knowledge.

Proteins, on the other hand, are known as the cells work-horses. Look anywhere in a cell and you will see proteins at work. Proteins are built in thousands of shapes and sizes, each performing a different function. Some simply take on different shapes - rods, nets, hollow spheres, tubes – to become a part of more complex micro-machinery in the cell. Some are molecular motors, using energy to rotate, or flex, or crawl. Many are chemical catalysts that perform chemical reactions to transform chemicals into needed product in a critical chain to ensure that the body gets all that is needed to sustain itself. With their wide potential for diversity, proteins are constructed to perform most of the everyday tasks of the cells. In fact human cells build around 30,000 different kinds of proteins to execute the diverse array of cellular level tasks. Proteins hence, exist for service, to bring about perfection at the level of the cell, are characterized by extreme diligence and perseverance, and so on. Being so, it can be seen that proteins are likely an emergence of Light's property of Presence.

Lipids, the third ubiquitously available molecular plan, are by themselves tiny molecules, but when grouped together form the largest structures of the cell. When placed in water, lipid molecules aggregate to form huge waterproof sheets. These sheets easily form boundaries at multiple levels and allow concentrated interactions and work to be performed within a cell. Hence, the nucleus and the mitochondria are contained within lipid-defined compartments. Similarly, each cell itself is contained within a lipid-defined boundary. Lipids are therefore promoters of relationship, of harmony in the cell, of nurturing the cell-level division of labor, of allowing specialization and uniqueness to emerge, hence perhaps of earlier forms of compassion and love, amongst related functionality. This function of harmonization suggests that lipids can therefore be thought of as an emergence of Light's property of Harmony.

Finally, Polysaccharides are long, often branched chains of sugar molecules. Sugars are covered with hydroxyl groups, which associate to form storage containers. As a result polysaccharides function as the storehouse of the cell's energy. In addition polysaccharides are also used to build some of the most durable biological structures. The stiff shell of insects, for example are made of long polysaccharides. Polysaccharides function to create energy, power, courage, strength thereby readying the cell for adventure, and so on. Providing energy and strength, polysaccharides can be thought of as an emergence of Light's property of Power at the cellular level.

Love

In our ongoing exploration of Light we have discovered how it's slowing done could have created the whole universe through the Big Bang. We have also seen how the nature of realities that may exist is also dependent on the way light moves. So it can be seen that there are these incredible, existential mysteries in Light.

Here we are going to explore one more mystery of Light – that of Love.

Let us consider again each of the layers of matter and life we just explored. We can see that four properties of Light – Power, Knowledge, Harmony, and Presence – have structured these layers. But in each of these cases it is not one of these, but all four working together that creates the foundation of the layer.

So at the electromagnetic field level, there is a sustainable wholeness that comes into place because the possible archetypes embedded in the electromagnetic spectrum – the knowledge aspect, exists as a range of frequencies that allows different energies to come into being – the power aspect, and that can create a range of particles – the presence aspect, that all travel at a speed that allows what is created to interact – the harmony aspect. Without the participation of each of these aspects subsequent layers of matter could not emerge. But such complete interaction of different aspects working together to create a whole different possibility is nothing other than Love.

So we see that the properties of Light that project themselves out for the purpose of creation are bound together in an external wholeness by Love, which has too to be implicit or contained in Light itself.

But exploring further, at the quantum level, it can similarly be said that quarks or leptons or bosons or the Higgs-boson by themselves would not have been able to create the foundation for the next layer of complexity in Light's journey. If they acted independently there would have only been a flurry of endless flows. It is only the combined action of the four held together in a tight embrace that is able to create the reality of the atom. But such a tight embrace of distinct actors or personalities that causes cohesive and unified action is nothing other than Love.

Hence it is Love that pulls together the possibilities projected by Light to create a unity that must exist for any next layer of complexity to emerge.

Continuing with the exploration of the layer of quantum particles, it is the coordinated action of bosons that allows the richness, the possibilities, the dynamism of quantum particle interaction, and all the myriad results of this to come into being. Gluons allow the nucleus to form and therefore allow the sustained codification of what an element will mean in the scheme of things to come into being. W and Z bosons allow the release of energy to be recycled or used for other purposes of interaction, through decay of heavier quarks and leptons. Photons as the carrier particle of the electromagnetic spectrum allow this field to be ever-present in this reality. Hypothetical gravitons would allow built-up objects to enter into a space-time relationship with each other: it bounds objects to behave in a unified way in a specified space and for a specified time. But all this holding together, and releasing, and recycling, and relating, this cohesiveness of the action of the force-carriers or bosons is only another instance of the action of Love. Any of these forces acting by themselves would not be able to create the foundation on which subsequent emergences are built.

Similarly, the complexity of molecules resulting in the diversity of possibility is only so because such a molecular foundation is built by Love: it is Love that causes a sustainable molecular-foundation constructed from Power, Knowledge, Harmony, and Presence type atoms to come into being. A universe created by one type of atom only would quickly be crushed under the weight of an ever more unsustainable imbalance.

Cells themselves require a balance of lipids, polysaccharides, proteins, and nucleic acids to meaningfully sustain themselves. Such balance is the sign that Love has left its stamp.

So where there was only Light, Love too was. Where Light has projected itself into ever-more crystallized forms of matter and life, there too Love must be. This is so because Light in its wholeness is inseparable. The properties of Light when projected therefore, still have to act as though one because that is the innate law of their being. This continued action of oneness in crystal form is what Love is, and it acts so that the properties of Knowledge, Power, Harmony, and Presence can more fully manifest what Light really is.

Sensations, Urges, Feelings, Thoughts

As human beings we also experience sensations, urges and desires and wills, feelings and emotions, and thought.

But could it be that these are also projections or emergences of the properties of Light?

Sensations are those things we experience with our senses. We see things, hear things, and smell things, taste things, can touch things. This ability to enter into relationship with objects through sensation is nothing other than a result of the emergence of Light's property of Presence. We become present to Presence through the device of sensation. Sensation can be thought of as the means by which this property of Light – Presence - molds or ingrains itself in us as human beings. Its potentiality, all which is contained in this aspect of Light, becomes available to us through the power of sensation.

So also the path of development of sensation becomes clearer when viewed in this way. When we see things, for instance, what are we seeing? Is it just the surface rendering of the play of matter, or do we see that the fullness of Light is still there, with all its potentiality and possibility, in the smallest thing we look

28

at? Do we see that the whole universe and more is present in all its fullness in the least thing that we easily ignore, or belittle, or loathe? When we touch things is it the seeming concreteness of the play of the particles or atoms or chains of molecules that we touch? Or is it the Love and Light and the vastness of all that IS that allows itself to be as a small corner that we touch so as to make infinity be felt by something so finite?

Urges and desires and wills are similarly a play of the emergence of Light's property of Power. In the mystery of focus, the vastness of Light has projected itself in us into an apparent smallness that is in reality everything that is. And this smallness is trying through urge and desire and will to connect viscerally or even intentionally to other smallnesses that similarly are nothing other than the fullness of Light projected into a small smorgasbord of selected function. So the urge or desire for food, or companionship, or of possession, or of climbing a peak, is nothing other than Light's compressed property of Power, trying to reach more of the fullness that it is through a fulfillment of the urge or desire or will that it masquerades as.

Feelings and emotions are a play of the emergence of Light's property of Harmony. Its instrument is the Heart and it generates an array of emotions that are an indication or active radar of whether we are moving toward or away from a reality of harmony, whether based on our small self or some larger Self of Light. Gradually, by navigating with these emotions and feelings we can get to a state where we always feel positive emotions which basically means we have more truly entered into relationship with some larger continent of Light.

Thoughts are a play of the emergence of Light's property of Knowledge. Through the thought we can become greater or conceptualize things greater or begun to enter into relationship with some things other than our small self. Thought allows us to connect to more "othernesses" or even the oneness of the reality of Light.

So we can see that Light or the properties of Light have seeded our very beings. We are in reality nothing other than something of Light. Our very means of interacting, of connecting, of knowing, of relationship, are nothing other than gifts from Light. So if we feel small, or isolated, or dejected, or unsure, it is only because we have allowed the reality of Light to recede somewhere in the background, and forget that that is what we really are.

Having laid the foundation of Light that created the universe, and then projects itself or emerges in a four-foldness of Presence, Power, Knowledge, and Harmony to create mater and life, we begin to explore some of the astounding implications of this:

- All is Light
- Light Mathematics – Creating the World with Light
- True Individuality

So we have seen that this universe is due to a play of light. Light slowed down and was able to exist in packets or quanta. In turn this allowed matter to be sustained, without dispersing right away, which has given concreteness to everything that we see around us.

But further, everything we see around us is nothing other than an emergence of the play of Light.

When we look at our hand it is in reality made up cells that comprise of nucleic acids, polysaccharides, lipids, and proteins. These are themselves forms of knowledge, power, harmony, and presence that are none other than a more dense functionality of the properties of Light.

But these molecular plans are themselves made from chains of molecules comprised primarily from oxygen, carbon, hydrogen, nitrogen, calcium, and phosphorus. The property of power – surfacing in hydrogen and calcium, and the property of knowledge – surfacing in oxygen, carbon, nitrogen, and phosphorous - exist in abundance in our bodies. But the trace elements manganese, iron, copper, zinc all derived from Light's property of presence – are there too and essential to the functioning of cells. And lanthanides that are a projection of Light's property of harmony are known to exist in the blood. So again we see that the whole functioning of the hand is dependent on Light within Light.

But looking in even deeper, every atom is itself made from the interaction of quarks, leptons, boson, and the Higgs-boson that are again emergences of Light's knowledge, power, harmony and presence properties.

So we see that our hand is itself a crystallization of several layers of Light.

And the Light did not diminish as it takes on these emergences. It is there in its fullness, just as the Ocean is present in its fullness in a wave or the spray that is flung from it. So our hand is all the fullness of creation and all the potentiality of creation present in the form of our hand.

But so then is everything else that we see around us, from the smallest to the largest. All is a tremendous play of Light and remembering That, the universe and all its creations take on a very different reality and meaning. For then we find that we are always with our Lover, our Illuminator, our Force, our Presence, our Harmonizer or Nurturer. There has never been any separation. The only separation has been because of the paucity or incompleteness of our vision.

Light Mathematics – Creating the World with Light

When we look at a tree or a flower or an animal there is often a distinctness that becomes apparent. Considering the plant kingdom for example, there is a huge array of medicinal properties that are known to exist in plants. These can in turn be used to address the spectrum of afflictions that surface in human bodies. In other words there is abundant 'function' that exists in the variety of plants. Similarly, without considering the individuality within a type of animal, one can see that the type of animal itself seems to embody a different function. Some animals easily become pets while others will always remain the lone predator. Some function best with distinct roles in a larger collectivity, and some just appear to be wiser.

One can see that there is an intricate combination and play of function that can be behind the million forms we see and experience daily.

In fact imagine the four properties of Light - Power, Knowledge, Presence, Harmony as being four massive streams with countless variations within each stream. The variations are consistent with the theme of the overarching property so that sub-properties, in a sense, such as courage, adventure, leadership, justice, upliftment and so on all belong to or exist in the stream of Power. Similarly, wisdom, law-making, subject-specific knowledge, and so on belong to the stream of Knowledge. Perfection, diligence, sincerity, perseverance, humility, and so on along exist in the stream of Presence. Mutuality, compassion, nurturing, relationship, collectivity, and so on, all exist in the stream of Harmony.

Now further imagine that these massive streams rush together and from their union countless seeds are formed. These seeds are such that one of the myriad elements dominates, and then there are possibly numerous other elements that combine with the main one, to create a plethora of unique seeds.

So there is a kind of mathematics of light at work such that the properties and their variations of light become the unique, infinite number of seeds that are the basis of all the infinite functionality and combination of functionality we see around us.

33

So a plant may be a projection of such a unique functional seed, or may be the result of a number of functions coming together. But since the seed is created in such realms of light, when it is projected into time and space, there are the dynamics of separation from the source that are more dominant, and the seed has to grow and assert what it is in its essence through the play of time and space again. So the inner or light-based creation has to create itself again in worlds apparently separated from light, and in their growth have to reestablish the connection with the larger continent of Light.

So all can be seen as this tremendous 'Light Mathematics' where seeds of uniqueness are created by properties of Light, and then projected into space-time worlds, which as a reminder are a result of the slowing down of Light, and then grow into their essence to establish again a concrete contact with the inner-world of light moving infinitely fast and from which it was created.

Truer Individuality

Given, then, that we are all in our core, projections of seeds formed in a vast continent of Light the question is how do we connect with that - the seed - consciously?

That is always there but it is covered by dynamics of the physical, the vital, and the mental. These too are formed from properties of Light, but since the light has been separated from its source these movements and dynamics are incomplete in their nature.

It is easy for these dynamics and movements to completely preoccupy and occupy us. So much so that we identify with them, believe that these movements are us entirely, and in the process become subject to them. But we need to remember that even in their separation and smallness - whether selfishness, myopia, fundamentalism, or any other ism - these movements are still light. All of us is Light, and this becomes the means by which we can loosen the apparent chains around us.

For in its essence these small movements are trying to extend into something other than what they are. By connecting the light in them with the larger Light this extension can yield something different than what is possible through any other means. And in the process the hold that these smaller movements have is diminished and one can begin to see or experience larger vistas of light.

Such a passage where one is always moving to larger vistas of light, more easily allows us to enter into a field of original seeds and it is so that we can begin to enter into conscious communion with our truer individuality.

The truer individuality itself is some mathematical function of the four properties of Light - Harmony, Power, Knowledge, and Presence, which in turn as we have already seen, are practically infinite sets of qualities related to one of the main properties of Light.

So it may be that one individual is primarily driven to connect people together, itself a variation of the property of Harmony. The individual may further want to do this by deeply understanding what makes these people each tick, itself a variation of the property of Knowledge. So the seed or the truer individuality of this person can be thought of as a mathematical function consisting of some element or property from the set of Harmony and some element or quality from the set of Knowledge, combined together with possibly different elements from the same or different sets, all with possibly different weights, but with the first weight of the need to connect people, being the strongest.

But connecting with this seed or truer individuality, and then always keeping that reality alive in oneself means that the play of light is altered, so that the deeper light from or closer to the continent of Light is allowed to participate more actively in the time-space world we practically live in, that is itself a result of the slowing down of the speed of light. So now this separated world is connecting to the continent of Light through truer individuality.

Truer individuality is allowing the separation between worlds or layers of light to dissolve and is causing this separated world to enter back into its nature of eternity or timelessness and spacelessness or infinity. Except that the individualities are now in concrete material form that then becomes a far more beautiful basis for continued projections and crystallizations from the continent of Light.

LIGHT MEDITATIONS

The knowledge of light can be used practically to begin to shift often-negative every-day experience. As examples the following practical situations and associated meditations are considered:

- Healing the Body
- Fullness of Light
- Meditation on Four-Foldness of Emotion
- Shifting the World with light

Our bodies are entirely a play of Light. Our bodies are made from cells and these cells are nothing other than a multi-layered structure emerging entirely from Light.

The first detectable layer is that of the electromagnetic field which structured as the range of waves, the gradient of energy, the ability to localize as particles, and moving at the speed c, is nothing other than a projection of Light's properties of Knowledge, Power, Presence, and Harmony respectively. Quantum particles are localization of this underlying field that similarly gather into quarks, leptons, bosons, and the Higgs-boson which are emergences of Knowledge, Power, Harmony, and Presence respectively. Bosons themselves that connect quantum particles together to create atoms surface as Gluons, W and Z bosons, Photons, and the hypothetical Graviton. But these emergences are themselves that of Knowledge, Power, Presence, and Harmony respectively. As a result quantum particles gather to create atoms, and all known atoms are similarly either of the s-Group or Power, p-Group or Knowledge, d-Group or Presence, f-Group or Harmony categories. All molecules are made from these known atoms and therefore are nothing other than combined functions of the infinite variation of the essential four properties of Light. All cells are made from molecules, and further, are themselves divided into Nucleic Acids, Polysaccharides, Lipids, and Proteins, which are themselves nothing other than emergences of Knowledge, Power, Harmony, and Presence, which as we know are the properties of Light.

So our entire bodies are made from light and are nothing but light. An itch or a pain or a swelling or any other dysfunction is simply a result of the cells acting in their forgetfulness of their truth. Their truth is that they are concrete projections and crystallizations from an eternal and infinite world of Light and that they can partake in that reality. It does not necessarily mean that they themselves will become eternal and infinite, just yet, but it means that always they can allow some sense or even concreteness of eternity and infinity to enter into the dysfunctions and begin to heal them.

This is energy, or information, or more appropriately Light Medicine, and it is the panacea for any bodily ills. What is required is that the contracted time and space dynamics of pain or ill are allowed to open to the truth or Light or expansive timelessness and spacelessness that exist always behind it. The separation needs to be annulled by allowing the dysfunction to achieve fruition by connecting to the vastness of Light. For in its reality the pain or dysfunction is nothing other than a visceral urge to move away from something in which the separation is becoming more and more acute.

So, focus on an area in the body that is in dysfunction. See it in as much detail as you can – the tissues or muscles or bone or ligaments. See further that these are all made from cells. Now see these cells as nothing other than a play of light – the essential properties of light are there in every cell. But go deeper yet and see that all the molecules and atoms and quantum particles and the very field that everything emerges from are nothing other than the play of Light. Now know that Light can never be separated, and further that every smallest quantum particle is nothing than the Full Play of Light in an apparently small form. Now keep seeing this complete fullness of light in all the minuteness in the area of dysfunction. Keep doing this. Keep imagining that the tissue, the cells, the molecules, the atoms, the quantum particles, the very field from which all these emerge is flooded with the fullness of Light in its infinity and eternity and with its full play of properties and vast number of qualities. THAT is all there always, and everything in the area of dysfunction is nothing other than selective functions that in their separation have forgotten that connection. Reconnect everything. Keep reconnecting it. See that this area is nothing other than that which has emerged from Light, and is nothing other than the full play of Light in a little space.

Fullness of Light

Much of our time is spent in feelings, or thoughts, or urges that keep us occupied with smallness. We get entwined with the movement of separation often becoming one with it.

But if we can see that any of these movements has all the Fullness of Light behind them then something of that fullness can come into the smallness and that can alter the consequence of the smallness. For the fullness is everywhere and in everything already. The smallness is just here and now in what we have been captivated by. Being everywhere and in everything already the consequence of opening to fullness can be entirely different.

It is not that our little urge, or our desire, or our little thought will be guaranteed to be fulfilled because we are somehow opening to the fullness, but rather that the fullness, if we can open to it will make happen what is best from the point of view of the fullness of light. And living in that knowledge of the best possible happening, or living in the feeling of the best possible happening, allows the littleness or smallness that may have captivated us to open more to the fullness, and a virtuous cycle is then put in place.

The more we can practice this through the day, with all our smallnesses, the more will the largeness of light fill the smallness and allow a completely different dynamic to take place.

The Fullness of Light slowed down to c and allowed this matter-based creation - a complex crystallization of light – to come into being. But further the Fullness can be in the smallness, and can make the apparently small matter-based creations become full creations with access to oneness, and all the possibilities and potentialities resident in that Fullness.

To think, to feel, to sense, to will, to desire are great opportunities, great gifts even that each of us is always endowed with, a thousand times a day, if we use them as such. For being in their essence nothing but emergences of the properties of Light, and having all the Fullness of Light resident or behind each of these divisions, means that these divided dynamics that seem to lead us further from Fullness, can become means to lead us instead back into Fullness, or rather can become a passage by which the Fullness finds one more means or one more hold to manifest more of itself in these smaller infinitely diverse matter-based creations.

So thinking that the Fullness of Light is right here, and seeing that the Fullness of Light is right here, and willing and desiring that the Fullness of Light is right here, and loving that Fullness of Light so that it becomes more and more real right here is what will make the right here full of the Fullness of Light.

Meditation on Four-Foldness of Emotion

Light is one. When light surfaces as a layer its components get separated and it is only the action of Love that bring about the unity in matter and life. This has to be kept in mind when meditating on emotions.

So, knowing that emotions are an emergence of Light's property of Harmony, it should be remembered that Light's other properties are probably close by and surfaced too in the emergence of the emotion, and bringing these into a balance or unity will help the emotion be resolved if it is a negative emotion. For all emotion - and focusing here on any negative emotion, usually has a trigger, and this may be some expectation of attachment caused by the way we think of ourselves or what is ours or what should be ours or how someone should have

39

interacted with us. So there is this thought component that will accompany the emotion, and this thought component can be seen to be a surfacing of Light's property of Knowledge. But then the emotion will usually also surface somewhere in our body or cause discomfort somewhere in our body. And this location in the body can be thought of as a surfacing of Light's property of Presence. And finally the entire complex will have a charge to it, a kind of a power that makes us aware that something is not quite right. This surfacing of the charge can be thought of as an emergence of Light's property of Power.

So we see that the emotion has several aspects to it and these are always there and that these aspects are none other that the components or properties of Light. These aspects can continue out of balance and the negative emotion, its effect on the body, the thought that gives it legitimacy, and the charge it is creating can continue forever unless Love is applied to bring these components into a balanced unity.

The application of love would mean that the imbalance is adjusted so that if it were a restrictive thought that is perpetuating the negative experience that this restrictive thought is replaced by another kind of more expansive thought. It would mean that the resulting charge is acknowledged and the impact on the body is acknowledged and holding the four-fold reconfigured emotion-complex in non-judgment, something else is allowed to surface instead.

Such a practice will likely take repetition before results are seen.

Shifting the World with light

As our muscle for imagining, thinking, seeing, feeling, urging, willing, loving Fullness of Light grows, it becomes easier to see, feel, imagine, urge, will, love Fullness of Light in circumstances and situations that are apparently disconnected from us. We could be sitting in San Francisco, and engage in such Light exercises in or on other parts of the world entirely.

But starting small, and with the self, of which there are already so many opportunities, and then moving to larger and more distant spheres is a good way to proceed. Because in so doing it is as though a kind of "lubrication" sets in, or the way Light flows in and around the small self has begun to change. Then it becomes easier to exercise the Light in the other, more distant, larger spheres.

So for example, pick a part of the world where you know there are parties in conflict because they are embedded in different points of view, for instance.

Now, even if you do not know these different points of view, or do not understand the subtleties they are expressing, still see that the thoughts, the feelings, the actions that are being exercised there are a surface rendering of the play of light, and also that these dynamics are separated from the Continent of Light and therefore are often in chaotic motion.

So see first, or will, or imagine that all the Fullness of Light is right there and in fact everything, all the smallnesses are just emergences of that or projections of that, and THAT is right there informing everything. Imagine, or see, or love the Light in such a way that it surrounds the entire area and let it flow into every aspect and let every aspect be filled with it, and let the whole area be filled with it, so that there is nothing there but Light.

Do this in such a way that there is no personal will, or urge, or thought favoring one outcome or resolution over another. Just do it so that the Light itself, which is already in everything, may resolve the situation from its own point of view. If you know what some of the players in your area of focus look like, imagine them too in that sphere of Light. See them, every part of them down to the smallest quantum particles being nothing other than a play of light, and offer again the whole situation to the Light that it may be best taken up by the Light from its point of view.

If you have a bias to imagine such things, rather than see such things, or will such things, whatever your bias that is fine. Follow the emerging or emergent property of Light that most easily comes to you. But also then as you feel the flow of Light, use the other faculties or the fullness of all the properties, as well as you able to, in the exercise of shifting this other part of the world. So use your thought of Light, your will for Light, your urge for Light, your sense of Light, your seeing of Light, your feelings for Light, your hearing of Light, your love for Light, to all become active in that sphere of your focus. If possible keep doing it until something in you feels that it is enough for now.

41

Light, and the speed with which it moves, as we have explored, will create a particular reality and a particular kind of a universe. This section explores some more subtleties related to this for those who would like to begin to explore such things in further detail:

- Necessity of Constant Speed of Light and Variable Time and Space
- Multiple Universes
- Waves, Fields, Blocks, and Quadrality
- Expanding and Contracting Universes

The constant speed of light, at a speed less than an infinite speed, allows a build up of energy at quantum levels, that in turn allows any properties or possibility in light to express itself materially. When light travels at the speed of c – 186,000 miles per second – then the result of that is the material universe as we see and experience it, also with its division of time into a past, a present, and a future. In fact, we could say that for the known universe, as we experience it, light had to travel at c for it to arise. The speed of light had to be constant else there would be a variable, fluid reality to matter, likely displaying barely forming islands subject to sudden disappearance, reappearance, continuing ad infinitum.

But also since the distance traveled by an object or a ray is the result of the speed it is traveling at for the time involved, this gives us an interesting insight that Einstein based his Theory of Relativity on. Since the speed of light is constant, and has to be for the universe to be in its observable stable condition, this means time and distance (or space) potentially can vary. So if an object is traveling very fast, then time and space are going to be experienced differently by it, as compared with an object that is traveling considerably slower: as the speed of an object approaches c time slows down and distance contracts.

So an object that manages to travel at a speed of c will in some sense partake of the reality as experienced by c. It will transcend the conventions of time and space that are set up because of c and experience these differently.

Multiple Universes

Light traveling at c appears to have created the universe as we experience it. But what if light travels at different speeds. We have already looked at light traveling at an infinite speed and imagined the omnipresent-omniscient-omnipotent-omniloving reality that will be created.

But what if light were to travel at a speed somewhere between c and infinity? Let us imagine that light is traveling way faster than c, say at a constant speed of K, but considerably lower than infinity. If we again imagine such light traveling at the quantum level and possibly emanating from the nucleus of some kind of atom, the question is how long before that initial impulse would be experienced at a small quantum-level distance from its point of origination from within the nucleus?

The time it would take would be a lot less than light traveling at c. In fact it may be so fast that it cannot be bounded materially but would be experienced in waves only. There is, relative to the case with c, not an allowing of the build up of energy to a point where the field so created can express itself in a different, material form. The existence that will result will therefore be much more wave-like and the universe would take that form.

Now imagine that light is traveling even faster than K, and closer to an infinite speed. The expression of light would instantaneously be in vaster fields or spheres of light and this would be the basis of a radically different existence or universe.

So, it is possible that there are many, many different kinds of universes all given reality by a different, though constant speed of light.

Particles, Waves, Fields, Blocks, and Quadrality

We have often heard of the duality of the reality of light. Light can be experienced as a particle or as a wave. But all this is telling us is that in fact light is traveling at different speeds, and is being experienced differently as a result of it. Our senses have been adapted to a reality that has been created by light traveling at a speed of c. Consequently so have all our thoughts and structures of belief of what is possible and what is not.

But why couldn't light travel at many different speeds simultaneously? If it does travel at multiple speeds then the resulting reality would appear to be different and it perhaps is fair to say that there are different kinds of particles, and different kinds of waves, and different kinds of fields, and different kinds of blocks that would be created simultaneously as a result of this behavior of light. But then philosophically, why shouldn't this be since Light is infinity and eternity and any manner of possibility is in it, and any manner of possibility can be projected or emerge from it.

But no matter how may kinds of particles, or how many kinds of waves, or how many kinds of fields, or how many kinds of blocks the point is that these many different kinds approximate into a quadrality – a particle-wave-field-block quadrality. So the particle-wave duality that is talked of now can be considered to be a subset of a particle-wave-field-quadrality.

44

But what does this quadrality mean, practically? It means that always and at every instant there is present in every point multiple ways that light behaves, and multiple realities that can be tapped into, or even multiple universes that intersect.

An Expanding and Contracting Universe

An expanding universe is the reality of our current universe. But it has also been proposed that the universe will contract at some point, and some have even proposed that such expansion and contraction are recurring and part of a cycle.

We know though that an expanding universe is the result of the Big Bang – that phenomenon when Light slowed down from an infinite speed to the speed of c, in order that some arrangement of possibility contained within it would project or emerge into a matter-based existence. We know further that implicit in the Big Bang and subsequently everything that is created, is the four properties of Light of Harmony, Power, Knowledge, and Presence. These are envisioned to be large sets with many nuanced elements in them that combine in infinite ways to create seeds from which all manner of uniqueness emerges in the universe.

In fact, we may say that so long as there continues to be an expansion of the universe this means that these active sets are in fact getting larger. That is, there is a more and more nuanced reality that is being created, which in turn means that the material world is continually opening to the larger worlds of Light behind and always present. And this is the meaning of creation. It is so there is more and more of the possibility in the four properties of Light that surfaces, or emerges, or crystallizes.

In the history of this universe we know now that the four essential properties at least emerged as the electromagnetic spectrum, then quantum particles, then atoms, and then also cells. But also the sets associated with each of the four properties has grown astronomically as each subsequent layer of the properties of light emerged so that, for instance, the set of Presence at the cell level, comprising of proteins, is massive since there are over 30,000 proteins in a single human cell.

With each subsequent layer of the emergence of the properties of light these sets have to continue to increase so that the infinite potentiality resident in Light can be realized.

If for some reason though there is focus on smallness, and myopia, and selfishness, or further separation from fullness, then the four sets cannot increase in size, uniqueness and diversity cannot meaningfully emerge, and the meaning of a creation has been defeated. In this case there is a contraction that begins, and the whole effort will need to wait for some other moment for another Big Bang and another cycle of possibility to emerge.

So expansion, so long as it continues means that the emergent vessels of Light have adapted in such a way that they are allowing more of the nuanced nature of the four properties to be become real and increase the effective size of the four sets. Contraction, on the other hand, means that the emergent vessels of Light are adapting in a way that causes less of the nuanced nature of the four properties to become real as a result of which the field of possibility correspondingly continues to diminish until time and space contract back into the timelessness and spacelessness from which they emerged.

This also gives insight into our universe being a function-based as opposed to a form-based universe. The functions derive from the combination of elements from the four sets that exist in Light. Such "functions" may have been glimpsed in the past as codified by historical mythologies that attribute different functions to different "macro-beings" or gods and goddesses. But these mythologies must have necessarily only glimpsed portions of the infinite sets and so long as we as receptacles of light continue to open to Light there can never be or will never be an end to the range of "functions" that will continue to become apparent.

Hopefully this section will continue to grow:
- Invitation to Continue the Collaboration
- Availability of Book
- Special Thanks

If the ideas in this short treatise appeal to you, or inspire you, or capture your imagination I invite you to creatively contribute in one or more of a number of ways. Here I list a few possibilities as starters:

- The treatise can be enriched with drawing and illustrations. These can be for children or adults and would create stand-alone books
- The treatise can be translated into different languages to share insights with larger audiences
- Further thinking and practical experimentation to do with any of the sciences referred to here whether to do with astrophysics, optics, quantum physics, electromagnetic theory, chemistry, biology, or any other fields or areas related to these. Working with active scientists and academicians in these fields will no doubt prove invaluable
- Further thinking and elaboration to do with the function-based mathematics referred to here. The original dissertation explores the initial creation of such a function-based mathematics. Working with real mathematicians will no doubt enrich the effort
- Further exploration into psychology of individuals as it relates to the expression of four-foldness of light
- Further exploration with organizational development practitioners to work through the implication of applying the light-framework explored here in practical terms
- Development of software tools to elaborate on one or more aspects covered in the treatise
- Construction of other meditations for healing at different levels or actualizing of potential that you would like to share
- Creation of workshops, seminars, conferences, living-room meetings to do with Light as it is explored here
- Creation of certifications for to-be light-coaches and light-facilitators
- Distribution of this content to people who would be interested in this material from a scientific, mathematical, organizational, psychological, or general perspective
- Translation of ideas here into a movie or expression onto other creative media

Please contact me if you would like to pursue or collaborate on any of these or any other idea at: Pravir.malik (at) deepordertechnologies.com

The original 350-page dissertation containing the derivation of 225 equations is currently available at the University of Pretoria repository, here:
https://repository.up.ac.za/handle/2263/62779?show=full

PRINT

US:
https://www.amazon.com/dp/0990357422
UK:
https://www.amazon.co.uk/dp/0990357422
Germany:
https://www.amazon.de/dp/0990357422
France:
https://www.amazon.fr/dp/0990357422
Spain:
https://www.amazon.es/dp/0990357422
Italy:
https://www.amazon.it/dp/0990357422
Japan:
https://www.amazon.co.jp/dp/0990357422

KINDLE

US:
https://www.amazon.com/dp/B076YR6N2B
UK:
https://www.amazon.co.uk/dp/B076YR6N2B
Germany:
https://www.amazon.de/dp/B076YR6N2B
France:
https://www.amazon.fr/dp/B076YR6N2B
Spain:
https://www.amazon.es/dp/B076YR6N2B
Italy:
https://www.amazon.it/dp/B076YR6N2B
Netherlands:
https://www.amazon.nl/dp/B076YR6N2B
Japan:
https://www.amazon.co.jp/dp/B076YR6N2B
Brazil:

https://www.amazon.com.br/dp/B076YR6N2B

Canada:

https://www.amazon.ca/dp/B076YR6N2B

Mexico:

https://www.amazon.com.mx/dp/B076YR6N2B

Australia:

https://www.amazon.com.au/dp/B076YR6N2B

India:

https://www.amazon.in/dp/B076YR6N2B

Special Thanks

- Tony Hsieh and John Robert Cornell for their comments around formation of matter at c which caused me to add further clarification
- Shivalik Bakshi for pointing out nomenclature with physical constants
- Dr. Eric Scerri for taking the time to offer some comments on the section on Atoms, which I have since revised based on these
- Rachael Brown for her continued enthusiasm for this book
- Kiki Bakshi and Deborah Davis for their insistence that I further edit the book
- Rajesh Krishnan for the back cover photograph

AUTHOR'S BOOKS

Early Books

1. The Flowering of Management
2. India's Contribution to Management

Fractal Series

1. Connecting Inner Power with Global Change: The Fractal Ladder
2. Redesigning the Stock Market: A Fractal Approach
3. The Flower Chronicles: A Radical Approach to Systems and Organizational Development
4. The Fractal Organization: Creating Enterprises of Tomorrow

Cosmology of Light Series

1. A Story of Light: A Simple Exploration of the Creation and Dynamics of this Universe and Others
2. Oceans of Innovation: The Mathematical Heart of Complex Systems
3. Emergence: A Mathematical Journey from the Big Bang to Sustainable Global Civilization
4. Quantum Certainty: A Mathematics of Natural and Sustainable Human History
5. Super-Matter: Functional Richness in an Expanding Universe
6. Cosmology of Light: A Mathematical Integration of Matter, Life, History & Civilization, Universe, and Self

Applications of Cosmology of Light Series

1. The Emperor's Quantum Computer: An Alternative Light-Centered Interpretation of Quanta, Superposition, Entanglement and the Computing that Arises from it
2. The Origins and Possibilities of Genetics: A Mathematical Exploration in a Cosmology of Light

www.ingramcontent.com/pod-product-compliance
Lightning Source LLC
Chambersburg PA
CBHW022051190326
41520CB00008B/778